SpringerBriefs in Applied Sciences and Technology

Computational Intelligence

Series Editor

Janusz Kacprzyk, Polish Academy of Sciences, Systems Research Institute, Warsaw, Poland

SpringerBriefs in Computational Intelligence are a series of slim high-quality publications encompassing the entire spectrum of Computational Intelligence. Featuring compact volumes of 50 to 125 pages (approximately 20,000-45,000 words), Briefs are shorter than a conventional book but longer than a journal article. Thus Briefs serve as timely, concise tools for students, researchers, and professionals.

More information about this subseries at http://www.springer.com/series/10618

Dhaval R. Bhojani · Vedvyas J. Dwivedi ·
Rohit M. Thanki

Hybrid Video Compression Standard

 Springer

Dhaval R. Bhojani
Electronics and Communication
Government Engineering College
Rajkot, Gujarat, India

Vedvyas J. Dwivedi
C. U. Shah University
Wadhwan City, Gujarat, India

Rohit M. Thanki
C. U. Shah University
Wadhwan City, Gujarat, India

ISSN 2191-530X   ISSN 2191-5318   (electronic)
SpringerBriefs in Applied Sciences and Technology
ISSN 2625-3704   ISSN 2625-3712   (electronic)
SpringerBriefs in Computational Intelligence
ISBN 978-981-15-0244-6   ISBN 978-981-15-0245-3   (eBook)
https://doi.org/10.1007/978-981-15-0245-3

This Springer imprint is published by the registered company Springer Nature Singapore Pte Ltd.
The registered company address is: 152 Beach Road, #21-01/04 Gateway East, Singapore 189721, Singapore

Preface

The usage of social media and sharing of multimedia content over the Internet are rapidly increasing in recent time. Various facilities such as sharing of videos, downloading and viewing them, and videoconferencing based on the Internet are routine in the modern lifestyle. These all applications are consuming high channel bandwidth and capacity, but most of the channels have limited bandwidth. Therefore, the biggest problem for transmission of a high amount of video content is that it cannot be transmitted over a small bandwidth channel. So, to overcome this problem, compression of video is necessary before transmitting it.

Video is a sequence of image frames with a large amount of redundancy between two frames. This redundancy may be subjective or statistical. The main aim of video compression techniques is to reduce this redundancy using different types of mathematical and information technology-related techniques. The output of these techniques is compressed video content with reduced redundancy than in original video content. Based on the applications, video compression techniques are divided into two types: "lossy" and "lossless." The main aim of lossless video compression technique is reducing the data in video content for sharing and storing without losing information in video content. While lossy video compression techniques do the same but lose information in video content, and the same concept is envisioned by various video standards such as Moving Picture Experts Group (MPEG) with different versions such as MPEG-1, MPEG-2, and MPEG-4. These standards are widely used for the application of video content transmission over a communication channel which has less bandwidth and limited storage capacity. The main challenges faced by these standards are no good trade-off between quality of video content, algorithm complexity, and achieved compression ratio. The latest MPEG-4 standard has good algorithm complexity and objective-based algorithm, while MPEG-2 has less algorithm complexity and subjective-based algorithm.

This book aims to provide basic video compression standards, particularly MPEG with its different versions. The detailed working of standard video codec with the help of figures is given in this book, so that readers can easily get the concept of this basic video standard. After that, the hybrid video codec is explained with details to achieve improved performance compared to standard video codec without including the computation complexity of the algorithm. Finally, the comparison of the obtained results for these standards is discussed.

Overview of the Book

In Chap. 1, basic information and properties of various video compression standards are briefly discussed. The rest of the book covers various video compression standards in Chaps. 2–4 along with technical background for video compression, working and experimental results of MPEG-2 and modified MPEG-2 standards. In Chap. 2, various terminologies such as color conversion, estimation of motion, transform coding, quantization, zigzag scanning, and entropy coding that are used in video compression are discussed. The working of the standard video codec is covered in Chap. 3. In this chapter, the basic concept of video encoder and decoder along with experimental results is demonstrated with the help of real-time video signal. The working of the hybrid video codec is covered in Chap. 4. Finally, Chap. 5 gives a comparison of the obtained results of these standards.

Features of the Book

- Basic information on video compression standards
- Detailed working of standard video codec for compression of the video signal
- Extensive discussion on a modified version of standard video codec
- Inclusion of results of video compression for real-time video signals

Acknowledgements

My task has been easier and the final version of the book is considerably better because of the help we have received. Acknowledging that help is itself a pleasure. I would extend many thanks to all persons who helped achieving the final version of this book. The authors are indebted to numerous colleagues for valuable suggestions during the entire period of the manuscript preparation. I would also like to thank the staff at Springer, in particular Aninda Bose, senior publishing editor/CS Springer, for their helpful guidance and encouragement during the creation of this book.

Rajkot, Gujarat, India   Dr. Dhaval R. Bhojani
Wadhwan City, Gujarat, India   Dr. Vedvyas J. Dwivedi
Rajkot, Gujarat, India   Dr. Rohit M. Thanki

Contents

About the Authors

Dr. Dhaval R. Bhojani is an Assistant Professor, Department of E.C. Engineering, Government Engineering College, Rajkot. He earned hist PhD in video compression from JJTU, Rajasthan in 2013. His areas or research are Image & video processing, industrial automation and embedded systems. He has published more than 20 research papers in refereed and indexed journals, and has participated in conferences at the international and national level. He has member of ISTE and Institute of Engineers, New Delhi, India.

Dr. Vedvyas J. Dwivedi is a Professor (Department of E.C. Engineering, Faculty of Technology and Engineering) and Vice-Provost of C. U. Shah University. He has guided 7 Ph.D. theses, 43 M. Tech. dissertations, examined 8 Ph.D. theses from Indian Universities, 4 Ph.D. Theses from the United States based Universities, published 137 research and review articles, delivered 38 expert/resource talks, authored/co-authored 10 books, completed 12 research/consultancy projects, published 8 patents, chaired 16 sessions in national/international conferences of IEEE, IETE, IE (I), ISTE and HEIs. His expertise and interest areas are wireless-satellite-mobile-optical-RF-Microwave systems, sensor-energy-signal technology. His videos on YouTube and www.cushahuniversity.ac.in are available. He has earned B.E (Electronics Engineering), M.E. (Electronics and Communication Engineering), and Ph.D. (Electronics and Communication Engineering).

Dr. Rohit M. Thanki received his Ph.D. in electronics and communication engineering from C. U. Shah University, M.E. in communication engineering from G. H. Patel College of Engineering and Technology and B.E. in electronics and communication engineering from Atmiya Institute of Technology and Science, India. He has more than 3 years of experience in academic and research. He has published 10 books with Springer and 1 book with CRC press. He has published 13 book chapters in edited books which are published by Elsevier, Springer, CRC press, and IGI Global. He has also published 20 research articles, out of these, 6 articles in SCI-indexed journals and 14 articles in Scopus indexed journals. He is a reviewer of renowned journals such as IEEE access, IEEE Consumer Electronics Magazine, IET Image Processing, IET Biometrics, Soft Computing, Imaging Science Journal, Signal Processing: Image Communication, and Computers & Electrical Engineering. His current research interests include Image Processing, Multimedia Security, Digital Watermarking, Artificial Intelligence, Medical Image Analysis, Biometrics, and Compressive Sensing.

List of Figures

List of Tables

Chapter 1
Introduction to Video Compression

Keywords Compression · MPEG · Video · PSNR

After the last of the 1940s, data compression has been important terminology and widely used in the field of information sharing and distribution. The technique related to data compression aims to reduce redundancy presented in the data [1]. The compression technique has two types of blocks such as encoder and decoder which performs encoding and decoding of data in such a way the reduce redundancy in the data. The basic model compression technique is shown in Fig. 1.1 which contains encoder or compressor, communication channel, and decoder or decompressor. The work of encoder or compressor is encoded and reduced the data. This compressed data may be transmitted through the communication channel and given as input to decoder or decompressor. The work of decoder or decompressor is to get original data from the compressed data. Here, the ratio between the data rate of the encoder and the data rate of the channel is referred to as the compression ratio. When working of the encoder is more complex than working of decoder, then the type of model is referred to as asymmetrical [2]. The motion picture expert group (MPEG) is working on this type of model.

The data compression has a property like that it transforms data into a string of symbols which would require lesser bits compared to the representation of original data. The types of data such as text, audio, image, and video are transmitted on any communication channel. Out of these data, the videos are used more and more in today life and people are sharing it using various applications and devices.

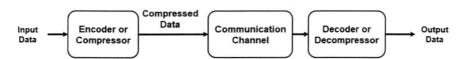

Fig. 1.1 Data compression model

© The Author(s), under exclusive license to Springer Nature Singapore Pte Ltd. 2020
D. R. Bhojani et al., *Hybrid Video Compression Standard*,
SpringerBriefs in Computational Intelligence,
https://doi.org/10.1007/978-981-15-0245-3_1

Table 1.1 Information on formats of various video signals

S.No.	Video format	Resolution	Require storage capacity (Mbps)
1	Quarter Common Intermediate Format (QCIF)	176 × 144	17.4
2	Common Intermediate Format (CIF)	352 × 288	69.61
3	Phase Alternating Line (PAL)	720 × 576	284.77
4	National Television System Committee (NTSC)	720 × 480	237.3
5	High Definition Television (HDTV)	1280 × 720	632.81

The digital videos can be considered as sequences of the image frame, and each frame has some amount of information in term of pixels. The value of these pixels defined the resolution of the video frame. Table 1.1 shows the resolution of various video formats with its required storage requirement [3–5]. These storage capacities require for storage of color video stream which has 30 frames per second capture rate.

As per referring to Table 1.1, it is indicated a small amount of video has required a large amount of data storage capacity. So, the large memory is required in any system which is built based on the video. Therefore, digital video compression algorithm represents video in a compressed manner with a smaller size compared to the original version and retains the quality of the original video in terms of perceptibility. It is working of the principle of perceptibility capacity of the human. The human eye cannot be identified as changes in chrominance components of color video. This property is widely used in many video compression standards which remove redundancy in video content to get its compressed version. The basic of video compression is discussed in the next subsection.

1.1 Basic of Video Compression

Video can be represented as a three-dimensional array with various digits, where the first two dimensions represent horizontal and vertical while the third dimension is the time coordinate. The video frame comprises of color pixels that correspond to a single time moment of a video. This frame is reference as video frame [6]. The video data has different types of redundancy in various domains such as spatial and temporal. Also, some other types of redundancies such as perceptual and statistical are associated with video data [7].

The similarities within a frame refer as spatial redundancy while similarities between two consecutive frames refer as temporal redundancy. The human eye cannot distinguish small changes in color components while it is clearly identifying

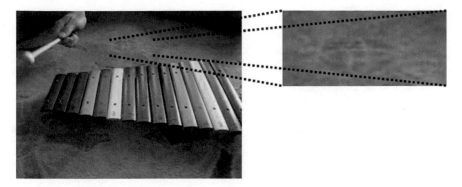

Fig. 1.2 Example of spatial redundancy

changes in brightness of similar color which has more spatial redundancy. The intra-frame coding is used reducing this type redundancy within the frame. The example of spatial redundancy is shown in Fig. 1.2. This type of redundancy can be reduced by various types of coding such as run-length, Huffman.

For reducing temporal redundancy, changes from one video frame to the next video frame identifies and using temporal compression encoded these changes. Here, temporal compression divides the whole video frame into nonoverlapping blocks and compression performs on them. These blocks refer to macroblocks. These blocks of one frame are then compared to the blocks of the next frame and encoder sends only differences between these blocks to reduce the redundancy. The example of temporal redundancy shows in Fig. 1.3 which shows a few frames of video having very minor changes from one frame to another frame. So, most of the temporal redundancy can be removed before transmission of it, which requires a small transmission data rate for the same video [4].

The perceptual redundancy refers that the details in video data which cannot perceive by the human eye. The information which can't perceive by human eye can be removed by the compression technique without affecting the quality of video data. The statistical redundancy is associated with different technique terminologies such as transform coefficients, vectors related to motion, codes. Therefore, a suitable chosen of this technique will get proper compression in video data.

Fig. 1.3 Example of temporal redundancy

1.2   Types and Need for Video Compression Techniques

The video compression techniques divide into two types such as lossy compression and lossless compression. In lossless compression, when the compressed video is decompressed, then resultant video has perfect matching with original video in the spatial domain. This type of compression is mainly removing spatial redundancy using intra-coding method. This type of techniques is rarely used due to less compression ratio [5, 6]. In lossy compression, when the compressed video is decompressed then resultant video has not given preface matching with original video in the spatial domain. This type of compression gives high compression ratio which predicts the changes in the frame using motion estimation and finds residue between two consecutive frames to reduced temporal redundancy in the frame.

Based on the coding method, compression divides into two types such as intra-coded and inter-coded. The intra-coding method works on three types of data such as x-direction and y-direction and sample value lie on location (x, y). The standard TV video signal contains high-frequency components in it due to a large number of details. Also, some information in small portion has similar pixel value which increases low-frequency components in it. The brightness of the video depends on zero frequency components. Therefore, suitable coding methods are required which work effectively in all types of frequencies in the video signal. The various signal transforms such as wavelet, discrete cosine transform (DCT) are used with the intra-coding method for suitable video compression. The inter-coding method works on similarities between consecutive frames of the video signal. This method sends only information about these similarities and reconstructed compression video using various differential coding.

The video compression technique has a trade-off between quality of the video, storage capacity, and cost of implementation. If the video compressed using the lossy technique, visible changes appear in the resultant video. Figure 1.4 shows the frame of the original video and its various compression version with different size of compression. From Fig. 1.4, it is seen that the compression increases storage capacity but decreases the quality of the video frame. After some rate of compression, the

(a)   **(b)**   **(c)**

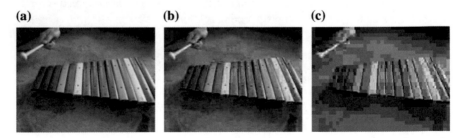

Fig. 1.4 Trade-off between quality of video and compression rate (CR). **a** Original frame (24.9 kB). **b** Frame with a low compression rate (3.65 kB). **c** Compressed frame with very CR (2.19 kB)

Table 1.2 Applications of video signal with its data rates

S. No.	Applications	Video resolution and frame per second (fps)	Uncompressed data rate	Compressed data rate
1	Videoconferencing	352 × 240 and 15	30.4 Mbps	64–768 kbps
2	CD-ROM digital video	352 × 240 and 30	60.8 Mbps	1.5–4 Mbps
3	TV broadcasting	720 × 480 and 30	248.8 Mbps	3–8 Mbps
4	HDTV	1280 × 720 and 60	1.33 Gbps	20 Mbps

quality of the video frame reduces such a way that the information of the frame cannot properly visible by the human eye.

Table 1.2 shows the applications of the video signal in real life with the required data rate for transmission of video signals and its compressed format. The table shows that due to compression, less amount of channel capacity required for transmission of this type of video signal and less memory for storage of it. This is one of the motivations behind development for various types of video compression standards.

1.3 Problems in Video Compression

The various types of problems are associated with video compression standard which is discussed below:

- The quality of the video signal is very sensitive, and compression method affects this quality parameter.
- The dropping of the video frame may happen during transmission of it. If the compression is not applied on individual framer, then some of the data may be lost after compression of it and proper reconstruction cannot be performed at the receiver side.
- The complexity of existing video compression standards is high.
- The video signal requires more bandwidth and so that it cannot be transmitted over a low bandwidth channel. Therefore, the compression of the video signal is required.

To overcome these problems in video compression, the hybrid video compression standard is discussed in this book which decreases the requirement of channel capacity and storage capacity while keeping video quality is higher.

1.4   Various Video Compression Standards

The various agencies worldwide are developed and defined various standards for video coding and compression. These standards are developed by two agencies which names are video coding expert group (VCEG) of telecommunication standardization sector of international telecommunication union (ITU-T) and moving picture expert group (MPEG) of the international organization for standardization (ISO/IEC). The standards developed by VCEG are called as H.26x and developed by MPEG are called as MPEG-x [7–11]. The history of these video compression standards is shown in Fig. 1.5. The information of these standards is given as below.

1.4.1   Motion JPEG

The video signal is nothing but a sequence of video frame which is treated as an image. Therefore, in this standard, taken the advantage of JPEG compression standards such as JPEG and JPEG 2000 and applied on sequences of images to achieve suitable quality and compression ratio. These standards refer as MJPEG and MJPEG 2000. The main disadvantage of these standards is that there are image compression standards and only applicable for sequences of images which cannot be treated as a video compression method [12].

1.4.2   H.26x

The details of various H.26x standards are given as below:

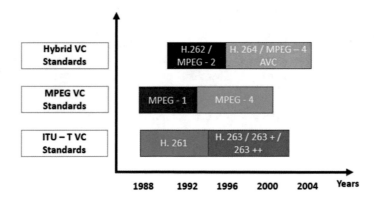

Fig. 1.5 History of video compression standards

- **H.261**: This standard was developed by the ITU-T group around 1990. This standard was designed for the transmission of video signals with a data rate of 64 kbps and its multiplying values. The frame of this standard can be divided into two types such as I (intra-coded) frame and P (predicted) frame. In I frame, coding of the frame is done without information of the previous frame where coding of the frame is done using the information of the previous frame in P frame. It is used motion-compensation temporal prediction to find similarities between two frames. This standard widely used in applications like video calling and videoconferencing.
- **H.263**: This standard was developed by the ITU-T group around 1995. This standard was designed for lower data rates. In this standard, the video frame was divided into number of macroblocks which have 16×16 blocks of the luminance component and 8×8 blocks of chrominance blocks. These blocks were encoded as intra- or inter-coding. In this standard, spatial redundancy is explored by DCT coding and temporal redundancy is explored by motion compensation. This standard encoded the signal with an accuracy of half-pixel and bidirectional coding. This standard provides a low bit rate which can be provided by standards like MPEG-1 and MPEG-2.
- **H.264**: This standard was developed with ITU-T together with the ISO/IEC group. This standard based on block-oriented motion compensation with providing a lower bit rate than existing standards such as H.263, MPEG-2, and MPEG-4. This standard provides flexibility in various applications related to networking and systems such as video broadcasting, storage of video in DVD, multimedia transmission and telephonic system which low and high bit rates are required. It widely used as a lossy compression standard but also provides lossless compression standard features. This standard provides 1.5 Mbps for transmission of digital satellite TV signal while MPEG-2 requires around 3.5 Mbps for the same signal.
- **H.265**: This standard refers to high-efficiency video coding standard and widely used for compression of the video signal. The aim behind designing this standard is that it provides higher coding capacity at a lower bit rate. It provides support to high-resolution video signal with maximum resolution up to 8192×4329. It provides a double compression ratio compared to other standards with same bit rate and video quality.

1.4.3 MPEG-x

The details of various MPEG-x standards are given as below:

- **MPEG-1**: This standard was developed by the MPEG group around 1991. This was the first standards developed by MPEG and accepted worldwide. It compressed video quality with a maximum allowed bit rate of 1.5 Mbits/s. The standard operates on data rate between 1 and 2 Mbit/s. This standard used motion estimation and compensation at 16×16 macroblocks with full-motion search as a coding method which was used by H.261 standard. Additionally, this standard use block-wise discrete cosine transform, quantization, and entropy coding. In addition to H.261, this standard added bidirectionally predicated frame and half-pixel motion search. Due to additional features, this standard improves the data rate of 1 Mbits/s compared to H.261 standard.

- **MPEG-2**: This standard was developed by the MPEG group around 1993. MPEG-2 standards mainly designed for achieved high compression ratio in the TV broad-casting signal. The targeted compression rate for this standard is 4–30 Mbps with providing high quality to the video signal. The additional feature in this standard is that it used the coding of interlace scanned picture which was for stability in the video signal and makes effectively transmission of it in various channels. The standard also introduced new feature such as video codecs' concept which pro-vides scalability in video signals and became a standard feature for another video compression standard.

- **MPEG-4**: This standard was developed by the MPEG group around 1993. MPEG-4 adopted new coding features such as motion estimation and compensation with variable block size, variable-length coding-based entropy coding, error removal capability for transmission of the video signal in compressed format. This standard supports the transmission of the video signal in lower bandwidth channel with high quality. The compression ratio of 100:1 is common for this standard. The standard mainly uses in various applications such as mobile communication where low bandwidth channels are used.

1.5   Applications of Video Compression Standards

The main application of video compression standards is to compressed video signals which are transmitted for TV broadcasting and required low storage or memory on the disk. The other possible applications of video compression are used in the broad-casting of UHF and VHF video signals, satellite TV broadcasting, manufacturing of VCD and DVD, videoconferencing, and transmission of high definition TV (HDTV) signals. Table 1.3 shows applications of various video compression standards with its required data rate.

1.6   Video Compression Model

The generalized block diagram for a video compression model is given in Fig. 1.6. This model consists of two types of blocks such as video encoder and video decoder. Here, a video frame encoded by encoder which converts the frame into a set of bits. After transmission of bits through the channel, the encoded bits are given to input to a decoder where the reconstructed frame is generated. If the channel is not error-free,

Table 1.3 Applications of video compression standards

S. No.	Standards	Applications	Data rate
1	MJPEG and MJPEG 2000	Sequences of still images	Not specified
2	H.261	Videoconferencing and telephony	64 kbps
3	MPEG-1	Storage of video on DVD/CD-ROM	1.5 Mbps
4	MPEG-2	TV broadcasting	>2 Mbps
5	H. 263	Videoconferencing	<33.6 kbps
6	MPEG-4	Motion object-based coding	Not specified
7	H.264	Videoconferencing, video surveillance, HDTV, HD-DVD	Not specified

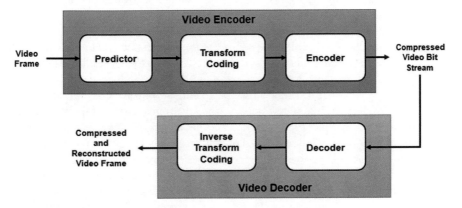

Fig. 1.6 Video compression model

then some distortion will appear in the reconstructed frame due to noise addition at the channel.

In Fig. 1.6, the encoder has different independent blocks such as predictor, transform coding, encoder while decoder has a decoder, inverse transform coding. The predictor has to identified various redundancy presented into the video frame. After that, according to presented redundancy in the frame, the appropriate signal transform is applied to the frame to get its transform coefficients. Finally, encoder removes coefficients which represent the redundancies in the frame. At video decoder, first, the decoder decodes the compressed video bitstream to get compressed transform coefficients of the video frame. After that, inverse transform coding is applied on it to get compressed and reconstructed video frame at the receiver side. The details working of video compression is also covered in Chap. 3.

1.7 Literature Survey on Video Compression Standards

In this section, the information on various existing video compression methods is discussed with advantages and disadvantages. The details are as follows.

Capon et al. [1] discussed run-length coding which was one of the entropy codings. This method was used at the last stage of the video encoder. It is used as source coding and encoded bits coming out of video encoder. This process reduces the number of required transmitted bits and reduces the channel capacity for transmission of it. Clarke et al. [6] gave the basic concept of compression of digital images and extension of it for the video signal. The author was explained image compression technique and applied the same technique for a sequence of images which can be treated as video. Seferidis et al. [13] discussed motion estimation and compensation which was an important method in any video compression standard. The authors also explained the block matching method and motion movement in fixed block size. The authors give information regarding motion vector which is important for the reconstruction of a video frame. Gharavi et al. [14] proposed some suggestion and modifications in the block-based matching method. They had described a hybrid motion estimation technique which provides good video quality.

Choi et al. [15] described motion vector quantization for the representation of encoded data in term of bits. They had described the transmission of motion vectors according to the direction of moving object. Here, motion vectors are differentiated according to moving objects, and based on this result, quantization and encoding of vectors are performed. Various researchers [16–20] were given coding based on a d-sequence method for error detection and correction in the bitstream of video for better reception of it at the receiver side. This method adds one addition bits in the bitstream of video which ales to check correct reception of video at the receiver side. Costa et al. [21] explained the working of MPEG-4 standard which provides scalability and flexibility in video bitstream using fine granular scalability (FGS) method. Authors have also explained the working of FGS method.

Luna et al. [22] explained the transmission of video signals using multiple channels. The main focus of authors in this approach is the better selection of coding parameters such as power, adaptation rate, and scheduling of transmission. Authors were given adaptive transmission algorithm which enhances efficiency and performance of transmission of video signals in any communication channels. Pereira [3] given a comparison of various video compression standards and applications of these standards for the real world. Farias et al. [4] gave no-reference video quality metric for video compression standard. This metric combined two-hybrid matrices which given overall degradation presented in compressed video. The metric also measured the blocking and blurring artifacts added into the video during transmission of it over a communication channel. Ouni et al. [5] presented a new signal transform which converts three-dimensional data into two-dimensional data. This transform explores temporal redundancy in the video frame for proper compression of it.

Chien et al. [23] discussed issued in the hardware implementation of video compression standards such as MPEG-4 codecs and H.264 codec. Authors described all aspects of hardware design and architecture of the system. Also, the authors gave chip for video compression standard like H.264 coded with the help of an example. Segall et al. [24] gave a survey on various compression standards with its advantages and limitations. Kondi et al. [25] proposed a new source–channel coding method using DCT for coding of the video stream. In this method, the optimization-based source coding and channel coding are used to reducing overall distortion in the video stream. Eisenberg et al. [26] studied and analyzed an optimization method for energy minimization required in video transmission and given some new directions for tackled with this problem. Experimental results of this method show that simultaneous changes in source coding and power of transmission provide good efficiency for video compression.

Wang et al. [27] proposed an optimal bit allocation method based on operational rate–distortion for video bitstream. This method allows the variable bits to assign to samples which provide maximum throughput. When the high data rate required for transmission of video, then this method provides a saving of bit rate which reduces the high amount of redundancy in it. Xiong et al. [8] gave a comparison study of two signal transforms such as DCT and DWT for compression of the video stream. The DWT overcomes the blocking artifacts presented in DCT-based video compression standard. Wiegand et al. [9] gave basic information of technical features of a video compression standard H.264/advanced video coding (AVC) and various applications of this standard in real life. Marpe et al. [10] discussed technical aspects of video compression standards such as H.265 and MPEG-4 with its important features. The author was also listed current applications where these standards are used. Huffman et al. [28] proposed source coding method which is one of the famous coding methods and used for encoding of video steam before transmission of it.

Li et al. [29] proposed vector quantization-based coding method for encoding of video steam. This method used for reducing temporal redundancy in color distribution in the video stream. Moorthy et al. [30] proposed lossless threshold concept for H.264 video compression standard. This method reduces psychovisual redundancy in the video stream. Fu et al. [31] explained working of motion estimation algorithm which improves the quality of compressed video stream. Thiesse et al. [32] gave the study of video compression and watermarking of video. Authors were discussed that how data embed in motion vectors of chroma and luminance components of the video stream. Suresh et al. [33] proposed ACC-DCT-based video compression technique for the 3D video signal. This technique reduced spatial and temporal redundancy which improves the efficiency of compression for the video signal.

Chen et al. [34] discussed bus congestion problem in hardware of video codec. Authors proposed a new algorithm which removes these problems and reduces requires memory for hardware without affecting the performance of the system. Alvarez et al. [35] beautifully explained optical flow concepts for estimation of motion vectors in the video stream. Authors have added Lucas–Kanade approach in motion estimation to achieve good consistency for compression of video steam. Sullivan et al. [36, 37] gave an overview of MPEG-2 and MPEG-4 with a new feature such as FRExt. Authors have proposed an interactive region of interest (IROI)-based wavelet transformation method for coding of the video stream. Cheung et al. [38]

highlighted difficult problems in the hardware implementation of the H.264/AVC standard. Jiang et al. [39] discussed the lossless video compression method for MPEG-2 standards which base for any video compression standard.

1.8   Points Cover in the Book

In this book, the basic concepts of video compression are discussed with its experimental results. The main points covered in the book are as below:

- **Video Codec**: The video compression and decompression used for the implementation of a video compression method are known as video codec. The stepwise working of video compression block gives video bitstream as output. This video bitstream is fed to video decompression block which reconstructed the original video stream. The process of compression includes conversion of color space, re-sampling of chrominance components, motion estimation and compensation of frame, transform coding of the resultant video frame, quantization process, selection of transform coefficients using zigzag scan, and entropy coding of selected transform coefficients to get compressed video bitstream. The video decompression blocks performed reverse processes of all video compression blocks.
- **Hybrid Transformation**: In this book, the novel approach for transform coding is discussed for compression of video frames. Here, a triple transformation-based method based on truncated singular value decomposition (TSVD), discrete wavelet transform (DWT), and discrete cosine transform (DCT) is introduced in MPEG-2-based video encoders. Due to the addition of this method, the new method provides more compression ratio for a video stream. The DCT-based MPEG-2 provides a compression ratio of up to 12% which is increased up to 5% by this newly proposed method.
- **Experimental Results and Comparative Analysis**: This book also shows the experimental results of MPEG-2 video compression standard and hybrid video compression standard for real-time video signal. The comparative analysis of these standards is discussed with the help of different video quality metrics.

References

1. Capon, J. (1959). A probabilistic model for run-length coding of pictures. *IRE Transactions on Information Theory, 5*(4), 157–163.
2. Watkinson, J. (2012). *The MPEG handbook*. New York: Routledge.
3. Pereira, F. (2011, May). Video compression: An evolving technology for better user experiences. In *2011 2nd National Conference on Telecommunications (CONATEL)* (pp. 1–6). IEEE.
4. Farias, M. C., Carvalho, M. M., Kussaba, H. T., & Noronha, B. H. (2011, June). A hybrid metric for digital video quality assessment. In *2011 IEEE International Symposium on Broadband Multimedia Systems and Broadcasting (BMSB)* (pp. 1–6). IEEE.

5. Ouni, T., Ayedi, W., & Abid, M. (2009, May). New low complexity DCT based video compression method. In *2009 International Conference on Telecommunications, ICT'09* (pp. 202–207). IEEE.
6. Clarke, R. J. (1995). *Digital compression of still images and video*. Cambridge: Academic Press Inc.
7. Ponlatha, S., & Sabeenian, R. S. (2013). Comparison of video compression standards. *International Journal of Computer and Electrical Engineering, 5*(6), 549–554.
8. Xiong, Z., Ramchandran, K., Orchard, M. T., & Zhang, Y. Q. (1999). A comparative study of DCT-and wavelet-based image coding. *IEEE Transactions on Circuits and Systems for Video Technology, 9*(5), 692–695.
9. Wiegand, T., Sullivan, G. J., Bjontegaard, G., & Luthra, A. (2003). Overview of the H. 264/AVC video coding standard. *IEEE Transactions on Circuits and Systems for Video Technology, 13*(7), 560–576.
10. Marpe, D., Wiegand, T., & Sullivan, G. J. (2006). The H. 264/MPEG4 advanced video coding standard and its applications. *Communications Magazine, IEEE, 44*(8), 134–143.
11. Furht, B. (1995). A survey of multimedia compression techniques and standards. *Part II: Video compression. Real-Time Imaging, 1*(5), 319–337.
12. ISO/IEC JTC 1. (2003). Advanced video coding. *ISO/IEC FDIS 14496-10, International Standard.*
13. Seferidis, V. E., & Ghanbari, M. (1993). General approach to block-matching motion estimation. *Optical Engineering, 32*(7), 1464–1474.
14. Gharavi, H., & Mills, M. (1990). Block matching motion estimation algorithms-new results. *IEEE Transactions on Circuits and Systems, 37*(5), 649–651.
15. Choi, W. Y., & Park, R. H. (1989). Motion vector coding with conditional transmission. *Signal Processing, 18*(3), 259–267.
16. Herlekar, S., & Kak, S. C. (2002). Performance analysis of a d-sequence based Direct Sequence CDMA system. *LSU report.*
17. Kak, S. C., & Chatterjee, A. (1981). On decimal sequences (Corresp.). *IEEE Transactions on Information Theory, 27*(5), 647–652.
18. Kak, S. C. (1987). Generating d-sequences. *Electronics Letters, 23*(5), 202–203.
19. Kak, S. C. (1987). New result on d-sequences. *Electronics Letters, 23*(12), 617.
20. Kak, S. C. (1985). Encryption and error-correction coding using D sequences. *IEEE Transactions on Computers, 34*(9), September.
21. Costa, C. E., Eisenberg, Y., Zhai, F., & Katsaggelos, A. K. (2004, June). Energy efficient wireless transmission of MPEG-4 fine granular scalable video. In *2004 IEEE International Conference on Communications* (Vol. 5, pp. 3096–3100). IEEE.
22. Luna, C. E., Eisenberg, Y., Berry, R., Pappas, T. N., & Katsaggelos, A. K. (2003). Joint source coding and data rate adaptation for energy efficient wireless video streaming. *IEEE Journal on Selected Areas in Communications, 21*(10), 1710–1720.
23. Chien, S., Huang, Y., Chen, C., Chen, H. H., & Chen, L. (2005). Hardware architecture design of video compression for multimedia communication systems. *IEEE Communications Magazine, 43*(8), 123.
24. Segall, C. A., & Katsaggelos, A. K. (2000, October). Pre-and post-processing algorithms for compressed video enhancement. In *Conference Record of the Thirty-Fourth Asilomar Conference on Signals, Systems and Computers* (Vol. 2, pp. 1369–1373). IEEE.
25. Kondi, L. P., Ishtiaq, F., & Katsaggelos, A. K. (2002). Joint source-channel coding for motion-compensated DCT-based SNR scalable video. *IEEE Transactions on Image Processing, 11*(9), 1043–1052.
26. Eisenberg, Y., Luna, C. E., Pappas, T. N., Berry, R., & Katsaggelos, A. K. (2002). Joint source coding and transmission power management for energy efficient wireless video communications. *IEEE Transactions on Circuits and Systems for Video Technology, 12*(6), 411–424.
27. Wang, H., Schuster, G. M., & Katsaggelos, A. K. (2003, September). Object-based video compression scheme with optimal bit allocation among shape, motion and texture. In *Proceedings 2003 International Conference on Image Processing. ICIP 2003* (Vol. 3, pp. III-785). IEEE.

28. Huffman, D. A. (1952). A method for the construction of minimum redundancy codes. *Proceedings of the IRE, 40*(9), 1098–1101.
29. Li, Z., & Katsaggelos, A. K. (2002). A color vector quantization-based video coder. In *Proceedings 2002 International Conference on Image Processing* (Vol. 3, pp. III-673). IEEE.
30. Moorthy, A. K., & Bovik, A. C. (2011, June). H. 264 visually lossless compressibility index: Psychophysics and algorithm design. In *IVMSP Workshop, 2011 IEEE 10th* (pp. 111–116). IEEE.
31. Fu, P., Xiong, H., & Yang, H. (2011, May). A motion estimation algorithm for educational video compression. In *2011 Workshop on Digital Media and Digital Content Management (DMDCM)* (pp. 257–260). IEEE.
32. Thiesse, J. M., Jung, J., & Antonini, M. (2010, October). Data hiding of motion information in chroma and luma samples for video compression. In *2010 IEEE International Workshop on Multimedia Signal Processing (MMSP)* (pp. 217–221). IEEE.
33. Suresh, G., Epsiba, P., Rajaram, D. M., & Sivanandam, D. S. A low complex scalable spatial adjacency ACC-DCT based video compression method. In *2010 Second International Conference on Computing, Communication and Networking Technologies*.
34. Chen, W. Y., Ding, L. F., Tsung, P. K., & Chen, L. G. (2008, June). Architecture design of high-performance embedded compression for high definition video coding. In *2008 IEEE International Conference on Multimedia and Expo* (pp. 825–828). IEEE.
35. Alvarez, L. D., Molina, R., & Katsaggelos, A. K. (2004, October). Motion estimation in high resolution image reconstruction from compressed video sequences. In *2004 International Conference on Image Processing, ICIP'04.* (Vol. 3, pp. 1795–1798). IEEE.
36. Sullivan, G. J., Topiwala, P. N., & Luthra, A. (2004, November). The H. 264/AVC advanced video coding standard: Overview and introduction to the fidelity range extensions. In *Optical Science and Technology, the SPIE 49th Annual Meeting* (pp. 454–474). International Society for Optics and Photonics.
37. Sullivan, G. J., & Wiegand, T. (2005). Video compression-from concepts to the H. 264/AVC standard. *Proceedings of the IEEE, 93*(1), 18–31.
38. Cheung, W. F., & Chan, Y. H. (2001, June). Improving MPEG-4 coding performance by jointly optimising compression and blocking effect elimination. In *IEE Proceedings-Vision, Image and Signal Processing* (Vol. 148, No. 3, pp. 194–201). IET.
39. Jiang, J., Xia, J., & Xiao, G. (2006). MPEG-2 based lossless video compression. *IEE Proceedings-Vision, Image and Signal Processing, 153*(2), 244–252.

Chapter 2
Technical Background

Keywords Color · Chroma · Entropy · Motion estimation and compensation · Transform coding

Any video signal can be represented by a three-dimensional array of color pixels where the first two values are represented horizontal and vertical direction known as spatial coordinates of the pictures and the third dimension is time coordinate. While a video frame consists of color pixels of one single moment of continuous video. The representation of the single video frame is similar to a digital image. The video signal has two types of redundancies such as spatial and temporal in it. The removing of redundancies in the video using two types such as removing similarities within a single video frame (known as spatial redundancy) and between two consecutive video frames (known as temporal redundancy). The various types of techniques and methodologies are used for removing these redundancies which are described in the next sections.

2.1 Color Space Conversion

Any color in the world has a combination of three primary colors such as red, green, and blue which is called RGB. The RGB color space is one of color space which can be used for the representation of video signal. The brightness (known as luminance) and color (known as chrominance) of any color image or video can be represented separately. The weighted sum of three primary colors gives the brightness of any image or video signal. It is also possible that this color can be converted into different color space such as YCbCr. The YCbCr color space is widely used in video compression. In this color space, the luminance coefficients (Y) of the frame are

© The Author(s), under exclusive license to Springer Nature Singapore Pte Ltd. 2020   15
D. R. Bhojani et al., *Hybrid Video Compression Standard*,
SpringerBriefs in Computational Intelligence,
https://doi.org/10.1007/978-981-15-0245-3_2

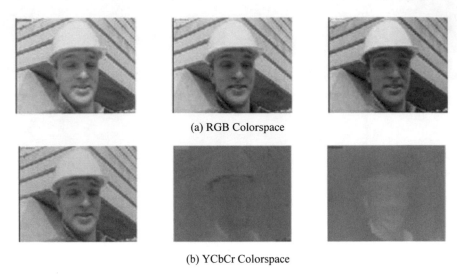

(a) RGB Colorspace

(b) YCbCr Colorspace

Fig. 2.1 Example of color space conversion

separated from the color information such as *Cb* and *Cr*. Here *Cb* gives the difference between the value of blue channel and reference while *Cr* gives the difference between values of the red channel and reference [1, 2].

Any RGB color signal can be converted into the YCbCr using the below equation:

$$Y = 0.299 \cdot R + 0.587 \cdot G + 0.114 \cdot B$$
$$Cb = 0.564 * (B - Y) + 0.5$$
$$Cr = 0.713 * (R - Y) + 0.5 \tag{2.1}$$

With the help of Eq. (2.2), the color signal in the YCbCr space can be converted into RGB space:

$$R = Y + 1.403 * (Cr - 0.5)$$
$$G = Y - 0.344 * (Cr - 0.5) - 0.714 * (Cb - 0.5)$$
$$B = Y + 1.773 * (Cb - 0.5) \tag{2.2}$$

The simple example of color space conversion is shown in Fig. 2.1

2.2 Re-sampling of Chrominance

The eye of human is very sensitive against any changes in the color information of signal compared to the luminance component. Therefore, the conversion of the signal

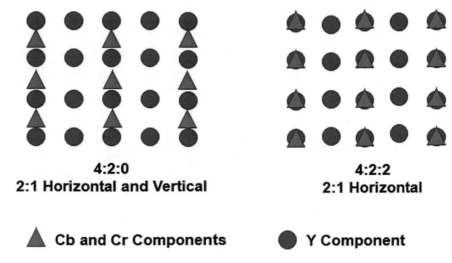

4:2:0
2:1 Horizontal and Vertical

4:2:2
2:1 Horizontal

▲ **Cb and Cr Components**   ● **Y Component**

Fig. 2.2 Example of re-sampling of chrominance components

from RGB space to YCbCr space is used in the compression of the video signal. In the re-sampling process, chrominance components Cb, Cr are re-sampling with a ratio of 4:2:2 by taking half pixels in horizontal direction or ration of 4:2:0 by taking half pixels in the vertical direction and horizontal direction. Due to this process, chrominance components reduce by 33% in the case of ratio 4:2:2 and 50% in the case of ratio 4:2:0 using the below equation. Thus, the re-sampling of chrominance components provides some amount of compression in the video signal. Figure 2.2 shows a simple example of the re-sampling of chrominance components.

$$4:2:2 = \frac{|y| + \frac{1}{2}|Cr| + \frac{1}{2}|Cb|}{|y| = |Cb| = |Cr|} = \frac{2}{3}$$

$$4:2:0 = \frac{|y| + \frac{1}{4}|Cr| + \frac{1}{4}|Cb|}{|y| = |Cb| = |Cr|} = \frac{1}{2} \tag{2.3}$$

2.3 Motion Estimation and Compensation

This is a very important process in video compression method. A video consists of a series of still images which are called as video frames. As taken two consecutive video frames which have a small delay and small changes in the pixel values of information, this is called temporal redundancy. This redundancy can be reduced by motion estimation and compensation process. Any group of the picture has three types of frames such as I frame, p frame, and B frame [3]. I frame is playing important

Forward Prediction

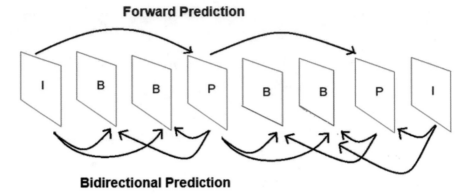

Bidirectional Prediction

Fig. 2.3 Prediction of frames in the Group of Pictures (GOP)

roles because it acts as a reference frame to other frames. This frame transmitted at the first and last position of a Group of Pictures. P frame is a predicated frame either come from the earlier I frame or from p frame. This frame can be reconstructed with the help of reference frame, and it has information like the difference between the reference frame and previously predicted frame. B frame refers to a bidirectional frame where one frame in the forward direction and second frame in a backward direction. P and B frames are also called an inter-coded frames while I frame is called an intra-coded frame. The prediction of P and B frames in a Group of Pictures is shown in Fig. 2.3.

The quality of compressed video stream is depending on the proper selection for sequences of frames in a Group of Pictures. The I frame provides good quality to video but not provides high compression ratio while P and B frames provide good compression ration but not given good quality video stream. Therefore, the trade-off between selection of these frames should be maintained to the obtained good performance of any compression standard. The normal sequence of the frame is used in practice as *IBBPBBPBBPBBIBBP...* [3].

The similarities between frames in GOP are found by the motion estimation and compensation algorithm. The similarity between frames is identified by motion vectors. For better performance of the technique, the estimation and compensation of motion vector should be good enough. This algorithm has a different process which is described as below:

- **Segmentation of Frame**: In this step, the frame is divided into macroblocks which different sizes such as 8×8 or 16×16 pixels. For the selection of macroblocks, two possible conditions occurred such as if the small size of blocks is selected, then more motion vectors are required for the representation of the frame. If the large size of blocks is selected, then matching of vectors is less. In practice, most of the standards are used 16×16 block size.

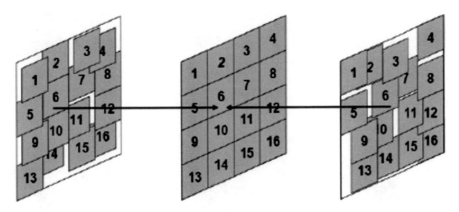

Fig. 2.4 Block matching process

- **Searching of Threshold**: It is a similarity level which used for comparison of similarity between two blocks. It is also used to decide where the difference between two blocks gives a difference in whole blocks or not.
- **Matching of Blocks**: This process arranged the position of macroblocks in a P frame to get actual p frame. This is a time-consuming process, and luminance components of the frame are used in this process. In this process, each block of the current frame is compared with all blocks of the previous frame. The important factor in this process is a search area. If the search area in the block is largely found by motion estimation, then it increases processing time. Therefore, rectangular-shaped macroblocks in the frame are seleted which were taken movement in both directions [4, 5]. The process of block matching is shown in Fig. 2.4 [4, 5].
- **Prediction of Error Coding**: In the encoding process, motion estimation is estimated and predicated the frame based on the reference frame. After that, this algorithm found a residual of the difference between the original frame and predicated frame [3]. The obtained residual referred as prediction error, and this error is coded for better transmission of the video signal. The simple process of finding prediction error is shown in Fig. 2.5 [3].
- **Motion Vector Coding**: The P and B frames are widely available in a group of pictures. These frames contain motion vectors which transmitted or stored in it. Therefore, the compression of these vectors is required to reduce the required number of bits. These vectors have a high correlation between each other. So, it can easily compress to these vectors [6].

2.4 Transformation Coding

After estimation and compensation of motion and residual in frames, frames are ready to transmission. So, these frames are converted into its frequency domain

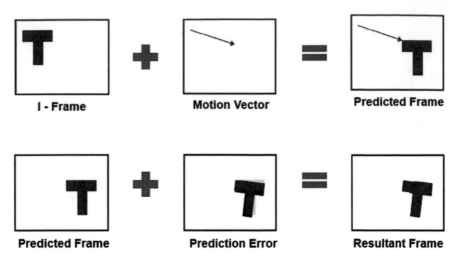

Fig. 2.5 Estimation of error in frame

using various types of signal transformations such as DCT, DWT, and SVD are used [3, 7]. The details of these transforms are given as below:

• **Discrete Cosine Transform (DCT)**

The representation of DCT of video frame is similar to fast Fourier transform (FFT). Here, the macroblocks with a size of 8 × 8 or 16 × 16 are taken as frequency components of the frame. The fixed size of macroblocks is taken due to the level of complexity and easy to implementation. The forward DCT can be defined using in below equation [8]:

$$F(u, v) = \frac{2}{N} C(u)C(v)$$
$$\sum_{x=0}^{N-1} \sum_{y=0}^{N-1} f(x, y) \cos\left(\frac{(2x + 1)u\pi}{2N}\right) \cos\left(\frac{(2x + 1)v\pi}{2N}\right) \qquad (2.4)$$

where $C(u)$, $C(v) = \frac{1}{\sqrt{2}}$ for $u, v = 0$; $C(u)$, $C(v) = 1$ for $u, v = $ else.

The calculation of DCT is very complex and time-consuming. If the size of the frame increases, then calculation time for DCT is also increased. That is the reason behind video compression algorithms and divides the frame into macroblock before calculation of DCT. The inverse DCT can be calculated using the below equation:

$$f(x, y) = \frac{2}{N} \sum_{u=0}^{N-1} \sum_{v=0}^{N-1} C(u)C(v)F(u, v) \cos$$
$$\left(\frac{(2x + 1)u\pi}{2N}\right)\left(\frac{(2y + 1)v\pi}{2N}\right) \qquad (2.5)$$

The basis of two-dimensional DCT and frequency distribution in the DCT block is shown in Fig. 2.6. The frequency distribution of DCT block indicated that components at leftmost position gives constant values and refer as low-frequency DCT coefficient or DC coefficients. As they move from left to right in the block, the frequency of DCT coefficients is increased. The portion highlights with the gray color indicated the DCT coefficients with mid-band frequency, and these coefficients are widely used in many applications such as watermarking and data hiding.

Fig. 2.6 **a** Basis of DCT.
b Frequency distribution in
DCT block

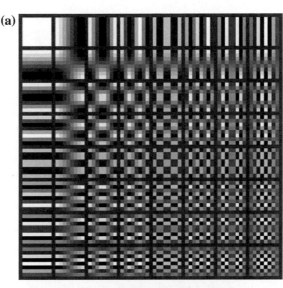

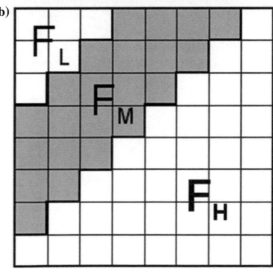

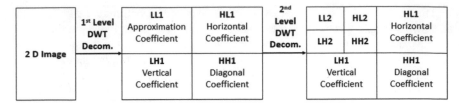

Fig. 2.7 Example of discrete wavelet transform (DWT)

• Discrete Wavelet Transform (DWT)

This is a famous signal transforms and works on sub-band coding. It converts video frame from spatial domain into frequency domain. It used multi-resolution technique, and the signal can be represented into different frequency sub-bands with different frequencies and resolution. In this wavelet decomposition, time scaling of the signal can be obtained using digital filters. Here, the signal passes through the low-pass filter and high-pass filter where the output of low-pass filter is approximation values of frame and detailed values of the frame as the output of high-pass filter. The output of DWT decomposition of any video frame with a resolution of $N \times N$ is in four different wavelet sub-bands such as LL, LH, HL and HH with a resolution of $N/2 \times N/2$. The basic example of a wavelet decomposition is shown in Fig. 2.7. DWT has excellent spatial—frequency representation property for video signal which becomes it is a very good candidate for video compression.

• Truncated Singular Value Decomposition (TSVD)

Singular value decomposition uses for minimization of storage of data and transmission of data. It is factorized any input data into three parts such as two orthonormal parts and one singular part. If SVD is applied to any video frame A, then it gives orthonormal parts such as U matrix and V matrix and one diagonal singular matrix S which gives eigenvalues of frame A.

$$A = USV^T$$

$$\text{Where, } U = [u_1, u_2, \ldots, u_n], S = \begin{bmatrix} \lambda_1 & 0 & \cdots & 0 \\ 0 & \lambda_2 & \cdots & 0 \\ \vdots & \vdots & \ddots & \vdots \\ 0 & 0 & \cdots & \lambda_n \end{bmatrix}, V = \begin{bmatrix} v_1 \\ v_2 \\ \vdots \\ v_n \end{bmatrix} \quad (2.6)$$

If the resolution of the frame is $M \times N$, then resolution of U matrix and V matrix is $N \times N$; resolution of S matrix is $M \times N$, but off-diagonal values of the matrix are zero. The singular value of S matrix must be fulfilled below equation:

$$\lambda_1 \geq \lambda_2 \geq \lambda_3 \geq \cdots \geq \lambda_n \geq 0 \quad (2.7)$$

(a)   **(b)**

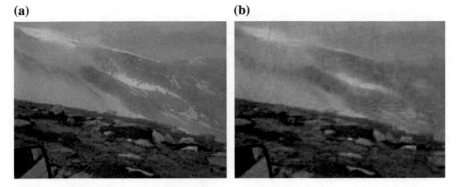

Fig. 2.8 a Original frame. **b** Truncated frame

The higher-ranked singular values have low values and fewer weights in the frame. Thus, these values eliminate in compression of a frame. If k number of singular values are used in compression, then less number of bits required for storage and transmission. The truncated SVD removes higher-ranked singular values and vectors which allows the only transmission of K number of singular values. The example of the application of truncated SVD on the frame is shown in Fig. 2.8.

2.5 Quantization and Zigzag Scanning

In the quantization process, the DCT coefficients of frames are divided into block size matrix which gives quantization. This matrix defined such a way that it eliminates high-frequency components to zero as human visual system less sensitive to these components. These components have fewer details regarding information of video frames. Due to this process, the degradation took places in the video frame but it can be controlled by the quantization matrix. The standard quantization matrix for video compression is shown in Fig. 2.9.

Fig. 2.9 Standard quantization matrix

16	11	10	16	29	40	53	68
12	12	16	23	32	43	56	71
14	13	21	28	37	48	61	76
20	23	28	35	44	55	68	83
29	32	37	44	53	64	77	92
40	43	48	55	64	75	88	103
53	56	61	68	77	88	101	116
68	71	76	83	92	103	116	131

Fig. 2.10 Example of
quantization process.
a Original coefficients.
b Process coefficients

(a)

10533	−96	−144	38	60	−47	−9	−14
−109	42	78	−61	−67	32	−5	6
−47	−44	85	−16	−36	−17	15	0
31	14	−92	−24	17	12	−36	0
−4	−42	5	9	−27	−36	37	1
38	−1	−89	−13	24	7	−38	−11
−48	−35	37	−6	−25	−24	22	9
−6	4	−34	−21	−2	4	−27	−6

(b)

64	−2	1	0	0	0	0	0
−5	−2	2	0	0	0	0	0
−5	−1	1	0	0	0	0	0
−2	−1	1	0	0	0	0	0
−1	0	0	0	0	0	0	0
0	0	0	0	0	0	0	0
0	0	0	0	0	0	0	0
0	0	0	0	0	0	0	0

The simple example of this process is shown in Fig. 2.10 below where image coefficients with the size of 8 × 8 blocks and values of same image coefficients after this process. It shows that most of the values of coefficients become zero after application of the quantization process on it. Due to this, more degradation and blocking artifacts appear in decoded video frames.

After the quantization of frequency coefficients, the zigzag scanning of coefficients is performed as shown in Fig. 2.11. This scanning performs such a way that arrangement of coefficients taken place in ascending order.

2.6 Entropy Coding

The two types of entropy coding such as run-length coding and variable Huffman coding are applied for the encoding of coefficients of video frames. In the run-length coding, the consecutive repeated numbers are represented by two numbers.

Fig. 2.11 Zigzag scanning
process

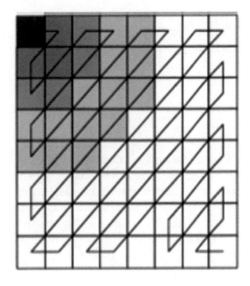

For example, if receive coefficients are (5, 8) as coded number, then it will be decoded as (8, 8, 8, 8, 8) and sequence of numbers [9]. The Huffman coding is a standard coding method which contains bit representation of all presented coefficients. This coding requires a smaller number of bits and used less frequent numbers are represented more numbers of bits. After this process, the video stream is ready to store on the optic disk or transmission of it via a communication channel.

2.7 Video Quality Matrices

In this section, information on various video quality matrices is given which are used for evaluation of the performance of any video compression standards or methods. The two types of matrices such as subjective and objective are used for this purpose. The most famous subjective measurement of degradation in the video frame is calculated using PSNR [10] which is given by the below equation. PSNR measures the ratio of maximum pixel values and the difference values between the original frame and compressed frame. It is measured on a logarithmic scale and given in dB.

$$PSNR = 10 * \log \frac{255^2}{MSE}$$

$$MSE = \frac{1}{M \times N} \sum_{x=1}^{M} \sum_{y=1}^{N} \{vf(x, y) - vf * (x, y)\}^2 \tag{2.8}$$

where vf is original video frame and $vf*$ is compressed video frame.

The similarity between the original frame and the compressed frame is calculated by another subjective measure like structural similarity index measure (SSIM) [11] which is given by the below equation:

$$\text{SSIM}(vf, vf*) = \frac{(2 \cdot vf \cdot vf * + c_1)(2 \cdot \sigma_{vfvf*} + c_2)}{(v\hat{f}^2 + v\hat{f}*^2 + c_1)(\sigma_{vf}^2 + \sigma_{vf*}^2 + c_2)} \tag{2.9}$$

where vf and $vf*$ are corresponding values for the original frame and compressed frame.

The SSIM measure the spatial similarity between the original frame and the compressed frame. The main difference between SSIM and PSNR or MSE is that SSIM considered the degradation as visible changes in structural information of the frame.

Another quality measure such as a mean sum of absolute difference (MSAD) is used for the quality of the compressed frame. This measure gives the difference between color components between the original frame and compressed frame, and it is given by the below equation:

$$\text{MSAD}(vf, vf*) = \frac{\sum_{i=1, j=1}^{M,N} |Vf(x, y) - Vf * (x, y)|}{MN} \tag{2.10}$$

The DCT-based video quality metric (VQM) gives mean distortion, and maximum distortion has taken place in compressed frame w. r. t. to the original frame and calculates using the below equation:

$$\text{VQM} = (\text{Mean\_Dist} + 0.005 * \text{Max\_Dist}) \tag{2.11}$$

The blurring beta gives a comparison of blurring artifact of the compressed frame with the original frame. The lower value of this metric indicated that most frames are blurred in compressed format. The blocking beta gives subjective blocking effect in a compressed frame with respect to the original frame. If the value of this matric is high indicated that more blocking artifact is presented in compressed video frame.

The compression ratio gives ratio between compressed video frame (which is bit steam outcomes from the encoder) and original video frame. It calculates using the below equation:

$$CR = \frac{DR\_vf*}{DR\_vf} \tag{2.12}$$

where DR_vf* is a data rate of the compressed frame and DR_vf is a data rate of the original frame.

References

1. Gonzalez, R., & Woods, R. (2008). *Digital image processing.* Delhi: Pearson Education India.
2. Gonzalez, R., Woods, R., & Eddins, L. (2009). *Digital image processing using MATLAB.* Delhi: TATA McGraw-Hill Education.
3. Moorthy, A. K., & Bovik, A. C. (2011, June). H. 264 visually lossless compressibility index: Psychophysics and algorithm design. In *2011 IEEE 10th IVMSP Workshop* (pp. 111–116). IEEE.
4. Seferidis, V. E., & Ghanbari, M. (1993). General approach to block-matching motion estimation. *Optical Engineering, 32*(7), 1464–1474.
5. Gharavi, H., & Mills, M. (1990). Block matching motion estimation algorithms-new results. *IEEE Transactions on Circuits and Systems, 37*(5), 649–651.
6. Choi, W. Y., & Park, R. H. (1989). Motion vector coding with conditional transmission. *Signal Processing, 18*(3), 259–267.
7. Li, Z., & Katsaggelos, A. K. (2002). A color vector quantization-based video coder. In *Proceedings of International Conference on Image Processing* (Vol. 3, pp. III-673). IEEE.
8. Jain, A. K. (1989). *Fundamentals of digital image processing* (pp. 150–153). Englewood Cliffs: Prentice-Hall.
9. Capon, J. (1959). A probabilistic model for run-length coding of pictures. *IRE Transactions on Information Theory, 5*(4), 157–163.
10. Petitcolas, F. (2000). Watermarking schemes evaluation. *IEEE Signal Processing Magazine, 17,* 58–64.
11. Wang, Z., & Bovik, A. (2004). A universal image quality index. *Journal of IEEE Signal Processing Letters, 9*(3), 84–88.

Chapter 3
Standard Video Codec

Keywords Decoder · Encoder · Quantization · Run-length

Video compression is a reversible process where video compressor encoded the video frame in a few numbers of bits. Throughout in video compression, the efficiency of the process for storage and transmission of data is monitored to achieve the desired compression of it. The inverse process of the encoder is called video decompressor where the compressed frame is decoded from the encoded data. The combination of the video encoder and video decoder is defined as video codec [1–8]. Here, the working of video codec [6] is discussed which was used in many video compression standards such as MPEG-2 and H.264 [1–8].

3.1 Video Compressor

The block diagram of the video compressor is shown in Fig. 3.1, and the working steps of it are discussed as below [6]:

Step 1 Video signal is divided into video frames and arranged in the GOP to get frame structure like I*BBPBBPBB*I….

Step 2 After obtaining video frames, all frames in RGB color space convert into YCbCr color space as per requirement of the human visual system (HVS). As per study, human eye is less sensitivity against chrominance components compared to luminance components. Therefore, to achieve a good quality of the video, more chrominance components and fewer luminance components are used in the compression process.

Step 3 After that, motion-compensated video frames are generated with the help of motion estimation and compensation algorithm. Here, frames are divided into macroblocks, and estimation of motion vectors is performed. According

© The Author(s), under exclusive license to Springer Nature Singapore Pte Ltd. 2020
D. R. Bhojani et al., *Hybrid Video Compression Standard*,
SpringerBriefs in Computational Intelligence,
https://doi.org/10.1007/978-981-15-0245-3_3

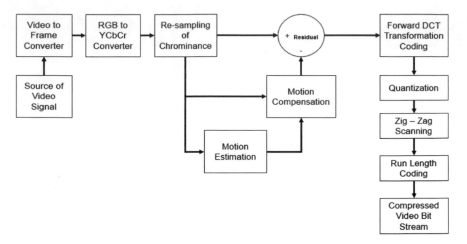

Fig. 3.1 Block diagram of video compressor

to motion vectors, the frames are predicted, and the residual between the predicted frame and original frame is referred to as the motion-compensated frame.

Step 4 The motion-compensated frame is converted into transform space using DCT-based transformation coding. Here, the residual frame is converted into its DCT frequency coefficients for further process.

Step 5 The quantization process is applied to DCT frequency coefficients of the residual frame. This process provides compression in the frame because it eliminates high DCT frequency coefficients of the frame which has less information.

Step 6 After that, zigzag scanning is applied to quantized DCT frequency coefficients in a block-wise manner. Due to this process, DCT frequency coefficients are rearranged in ascending order of frequency where low-frequency DCT coefficients are very important. These coefficients have more visual information of the video frame.

Step 7 After zigzag scanning, run-length coding is applied to these DCT frequency coefficients which are reducing the required number of coefficients to be transmitted. This coding reduces the required data rate for transmission of the frame. This coding is a subtype of Huffman coding.

3.2 Video Decompressor

The block diagram of video decompressor is shown in Fig. 3.2, and the working steps of it are discussed as below [6]:

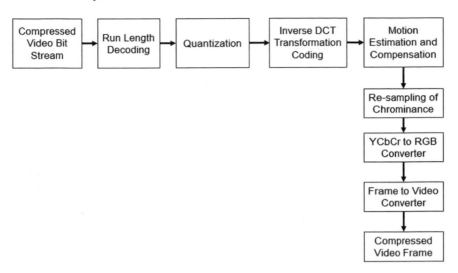

Fig. 3.2 Block diagram of video decompressor

Step 1 At the decompressor, the compressed bitstream is received as input.

Step 2 After receiving the data bits, the run-length decoding is applied on it to get quantized DCT coefficients of the frame.

Step 3 Then these DCT coefficients are de-quantized using quantization matrix which is generated at video encoder side to get decoded DCT coefficients of the frame.

Step 4 The inverse DCT is applied to decoded DCT coefficients to get motion estimated video frames.

Step 5 After that, the Group of Pictures (GOP) of the compressed frame is obtained from the motion estimated video frames with the help of motion estimation and compensation.

Step 6 Then re-sampling of chrominance components is applied on these compressed frames to get actual value of chrominance components of frames.

Step 7 Finally, inverse color space conversion such as YCbCr to RGB is applied to compressed frames to get compressed frames in RGB color space.

3.3 Experimental Results of Standard Video Codec

The implementation of video encoder and decoder is done with the help of MATLAB software. For the purpose of testing and analysis, different video signals are taken from Video Test media [9] and MATLAB test dictionaries which have different motions, colors, etc. For analysis purpose, the resolution of the video signal is set as 128 × 128 pixels and 30 frames per second (fps). The quality metrics are calculated with the help of the MSU Video Quality Measurement Tool 3.0 [10]. This tool measures all video quality metrics which are mentioned in Chap. 2, for evaluation and performance checking of video encoder and decoder. The compressed video frames along with measured quality matrices using explained video codec are given as below:

3.3.1 Results for Foreman Video Signal

The original foreman video frames and compressed foreman video frames are shown in Figs. 3.3 and 3.4, respectively. Here, the data rate of the video signal is 30 fps.

According to the frame rate in this test signal is 30 frames per second, so that 30 frames are given in compressed format. The compression ratio for this video signal is calculated using the below equation where compressed data rate is obtained after application of video encoder to the video signal.

$$CR_{Foreman} = \frac{((21856 + 23367 + 21869) * 2) + 73728 + 54}{128 * 128 * 3 * 30} = 0.114 \qquad (3.1)$$

In Eq. (3.1), the numerator values are obtained after obtaining final transmission coefficients after the process of run-length coding, motion vectors estimation on the video frame while denominator value is given the size of video signal 128×128 with 1-s duration for 30 frames. The average values of subjective quality matrices for foreman video signal are summarized in Table 3.1, while values of quality matrices for each frame are shown in Fig. 3.5.

3.3.2 Results for Vipman Video Signal

The original foreman video frames and compressed foreman video frames are shown in Figs. 3.6 and 3.7, respectively. Here, data rate of video signal is 30 fps.

According to the frame rate in this test is 30 frames per second, so that 30 frames are given in compressed format. The compression ratio for this video signal is calculated using the below equation where compressed data rate is obtained after application of video encoder to the video signal.

$$CR_{vipman} = \frac{((12158 + 12803 + 15429) * 2) + 73728 + 54}{128 * 128 * 3 * 30} = 0.0867 \qquad (3.2)$$

Fig. 3.3 Original foreman video frames

In Eq. (3.2), the numerator values are obtained after obtaining final transmission coefficients after the process of run-length coding, motion vectors estimation on the video frame while denominator value is given the size of video signal 128×128 with 1-s duration for 30 frames. The average values of subjective quality matrices for foreman video signal are summarized in Table 3.2, while values of quality matrices for each frame are shown in Fig. 3.8.

Fig. 3.4 Compressed foreman video frames for standard video codec

Table 3.1 Quality matrices for compressed foreman video signal for standard video codec

PSNR (dB)	MSE	SSIM	MSAD	VQM	Blocking beta		Blurring beta	
					Original frames	Compressed frames	Original frames	Compressed frames
22.65	353.47	0.7366	10.787	4.5864	7.15	7.914	24.52	26.808

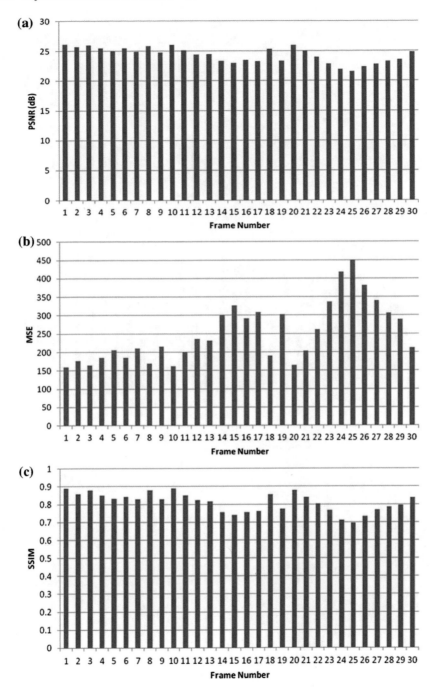

Fig. 3.5 Quality metrices for each frames of foreman video signal for standard video codec. **a** PSNR values, **b** MSE values, **c** SSIM values, **d** MSAD values, **e** VQM values, **f** blocking beta values, and **g** blurring beta values

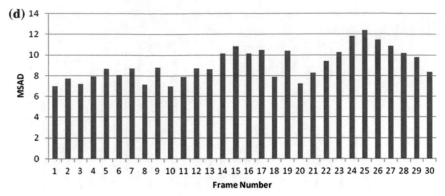

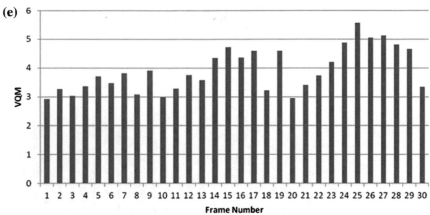

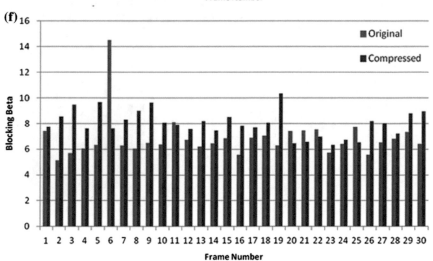

Fig. 3.5 (continued)

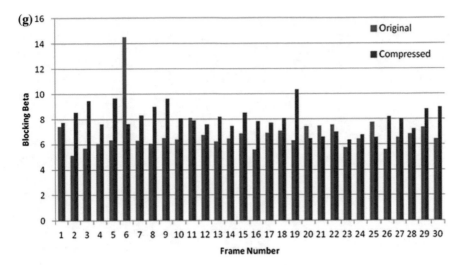

Fig. 3.5 (continued)

3.4 Summary of Chapter

The working of standard video compressor and decompressor is discussed in this chapter. The results of compressed video signals using this video codec show that the average compression ratio is around 8% to 15%. This compression ratio is enough for various applications such as TV broadcasting using cable wire or satellite and storage of video signal in CD-ROM/DVD. But when video signals are transmitted for the mobile network using the Internet, then high compression ratio is required. For achieving a high compression ratio, the modification in video codec is required. The one condition such as complexity of algorithm must be considered when the modification has taken place in video codec. Also, it provides a high compression ratio without affecting the quality of the video signal and can be used for video transmission in mobile networks.

Fig. 3.6 Original vipman video frames

Fig. 3.7 Compressed vipman video frames for standard video codec

Table 3.2 Quality matrices for compressed vipman video signal for standard video codec

PSNR (dB)	MSE	SSIM	MSAD	VQM	Blocking beta		Blurring beta	
					Original frames	Compressed frames	Original frames	Compressed frames
21.58	451.54	0.8139	9.124	5.3778	15.007	13.7187	15.757	18.577

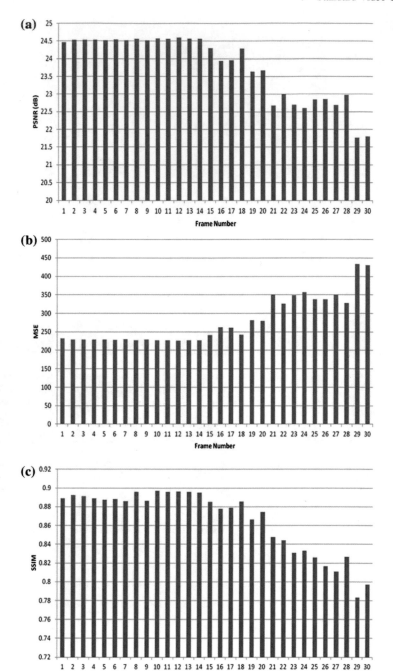

Fig. 3.8 Quality metrics for each frames of vipman video signal for standard video codec. **a** PSNR values, **b** MSE values, **c** SSIM values, **d** MSAD values, **e** VQM values, **f** blocking beta values, and **g** blurring beta values

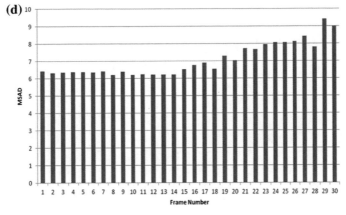

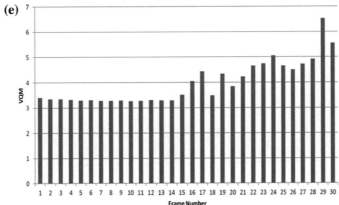

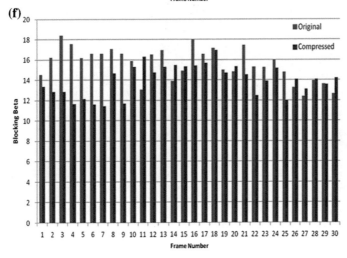

Fig. 3.8 (continued)

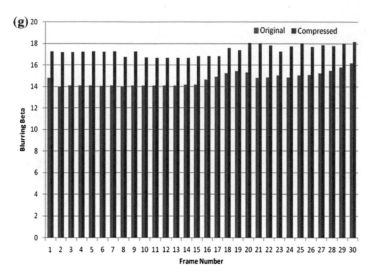

Fig. 3.8 (continued)

References

1. Capon, J. (1959). A probabilistic model for run-length coding of pictures. *IRE Transactions on Information Theory, 5*(4), 157–163.
2. Seferidis, V. E., & Ghanbari, M. (1993). General approach to block-matching motion estimation. *Optical Engineering, 32*(7), 1464–1474.
3. Gharavi, H., & Mills, M. (1990). Blockmatching motion estimation algorithms-new results. *IEEE Transactions on Circuits and Systems, 37*(5), 649–651.
4. Choi, W. Y., & Park, R. H. (1989). Motion vector coding with conditional transmission. *Signal Processing, 18*(3), 259–267.
5. Costa, C. E., Eisenberg, Y., Zhai, F., & Katsaggelos, A. K. (2004, June). Energy efficient wireless transmission of MPEG-4 fine granular scalable video. In *2004 IEEE International Conference on Communications* (Vol. 5, pp. 3096–3100). New York: IEEE.
6. Pereira, F. (2011, May). Video compression: An evolving technology for better user experiences. In *2011 2nd National Conference on Telecommunications (CONATEL)* (pp. 1–6). New York: IEEE.
7. Li, Z., & Katsaggelos, A. K. (2002). A color vector quantization-based video coder. In *Proceedings. 2002 International Conference on Image Processing* (Vol. 3, pp. III-673). New York: IEEE.
8. Moorthy, A. K., & Bovik, A. C. (2011, June). H. 264 visually lossless compressibility index: Psychophysics and algorithm design. In *2011 IEEE 10th IVMSP Workshop* (pp. 111–116). New York: IEEE.
9. Standard Video Signals. Weblink: https://media.xiph.org/video/derf/. Last Access: June 2019.
10. MSU Video Quality Measurement Tool. Weblink: http://compression.ru/video/quality_measure/vqmt_download.html. Last Access: June 2019.

Chapter 4
Hybrid Video Codec

Keywords Decoder · Encoder · Quantization · Run-length

As per results showed in the previous chapter, it is indicated that the standard video codec provides compression ratio in the range of 8–15%. If a user wants high compression ratio, then he or she moves to other video compression standards such as MPEG-4 which gives compression ratio around 5%, but this standard has more computational complexity and increases the processing time. So that, keeping in mind the computational complexity and high compression ratio, some modification can be done in a standard video codec which provides higher compression ratio by affecting the quality of the video signal. Therefore, in this chapter, hybrid video codec is discussed which provides high compression ratio. In this codec, some modification is done in transformation coding. In standard codec, DCT-based transformation coding is used, while in this hybrid codec, hybridization of truncated SVD and discrete wavelet transform (DWT) is used along with DCT-based transform coding. Here, first, truncated SVD is applied to residual video frame to obtain singular values of it. Then, single-level two-dimensional DWT is applied on the truncated frame which gives four wavelet sub-bands such as LL, LH, HL, and HH of the truncated frame. Finally, block-wise DCT-based transform coding is applied to LL wavelet sub-bands of the truncated frame. Other processes in hybrid codec are similar to the standard codec. The working of hybrid video compressor and decompressor is discussed in next subsections.

4.1 Hybrid Video Compressor

The block diagram of the hybrid video compressor is shown in Fig. 4.1, and the working steps of it are discussed as below:

© The Author(s), under exclusive license to Springer Nature Singapore Pte Ltd. 2020   43
D. R. Bhojani et al., *Hybrid Video Compression Standard*,
SpringerBriefs in Computational Intelligence,
https://doi.org/10.1007/978-981-15-0245-3_4

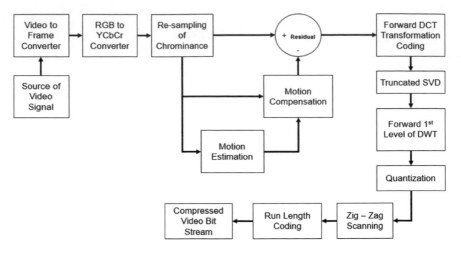

Fig. 4.1 Block diagram of hybrid video compressor

Step 1 Video signal is divided into video frames and arranged in the GOP to get frame structure like I*BBPBBPBB*I….

Step 2 After obtaining video frames, all frames in RGB color space converts into YCbCr color space as per the requirement of human visual system (HVS). As per study, the human eye is less sensitivity against chrominance components compared to luminance components. Therefore, to achieve a good quality of the video, more chrominance components and fewer luminance components are used in the compression process.

Step 3 After that, motion-compensated video frames are generated with the help of motion estimation and compensation algorithm. Here, frames are divided into macroblocks, and estimation of motion vectors is performed. According to motion vectors, the frames are predicted, and the residual between the predicted frame and original frame is referred to as the motion-compensated frame.

Step 4 The motion-compensated frame is converted into transform space using transformation coding. Here, the residual frame is converted into its hybrid frequency coefficients for further process.

Step 5 Truncated SVD is applied to the residual frame to reduce the size of it. Then first-level DWT decomposition is applied to the truncated residual frame to get its wavelet frequency coefficients such as LL, LH, HL, and HH.

Step 6 The forward DCT is applied to LL wavelet frequency coefficients of the frame to get its hybrid frequency coefficients.

Step 7 The quantization process is applied to hybrid frequency coefficients of the residual frame. This process provides compression in the frame because it eliminates high hybrid frequency coefficients of the frame which have less information.

Step 8 After that, zigzag scanning is applied to quantized hybrid frequency coefficients in a block-wise manner. Due to this process, frequency coefficients are rearranged in ascending order of frequency where low hybrid frequency coefficients are very important. These coefficients have more information.

Step 9 After zigzag scanning, run-length coding is applied to these hybrid frequency coefficients which are reducing the required number of coefficients to be transmitted. This coding reduces the required data rate for transmission of the frame. This coding is a subtype of Huffman coding.

4.2 Hybrid Video Decompressor

The block diagram of hybrid video decompressor is shown in Fig. 4.2, and the working steps of it are discussed as below:

Step 1 At the decompressor, the compressed bitstream is received as input.

Step 2 After the receiving of data bits, the run-length decoding is applied on it to get quantized hybrid frequency coefficients of the frame.

Step 3 Then these hybrid frequency coefficients are de-quantized using quantization matrix which is generated at video encoder side to get decoded hybrid frequency coefficients of the frame.

Step 4 First, inverse truncated SVD is applied on decoded hybrid frequency coefficients to obtain LL wavelet sub-band coefficients of the frame. After that, inverse first-level DWT is applied to obtain LL wavelet sub-band along original wavelet sub-bands LH, HL, and HH to get decoded DCT coefficients of the frame. Finally, the inverse DCT is applied to decoded DCT coefficients to get motion estimated video frames.

Step 5 After that, the Group of Pictures (GOP) of the compressed frame is obtained from the motion estimated video frames with the help of motion estimation and compensation.

Step 6 Then re-sampling of chrominance components is applied on these compressed frames to get actual value of chrominance components of frames.

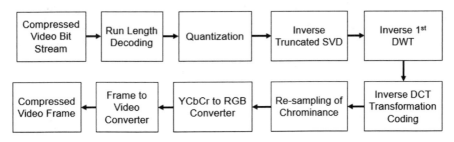

Fig. 4.2 Block diagram of hybrid video decompressor

Step 7 Finally, inverse color space conversion such as YCbCr to RGB is applied to compressed frames to get compressed frames in RGB color space.

4.3 Experimental Results of Hybrid Video Codec

The implementation of hybrid video encoder and decoder is done with the help of MATLAB software. For the purpose of testing and analysis, different video signals are taken from Video Test media [1] and MATLAB test dictionaries which have different motions, colors, etc. For analysis purpose, the resolution of the video signal is set as 128 × 128 pixels and 30 frames per second (fps). The quality metrics are calculated with the help of the MSU Video Quality Measurement Tool 3.0 [2]. This tool measures all video quality metrics which are mentioned in Chap. 2, for evaluation and performance checking of video encoder and decoder. The compressed video frames along with measured quality matrices using explained video codec are given as below.

4.3.1 Results for Foreman Video Signal

The original foreman video frames and compressed foreman video frames are shown in Figs. 4.3 and 4.4, respectively. Here, the data rate of the video signal is 30 fps.

According to the frame rate in this test signal is 30 frames per second, so that 30 frames are given in compressed format. The compression ratio for this video signal is calculated using the below equation where compressed data rate is obtained after application of video encoder to the video signal.

$$CR_{Foreman} = \frac{((8106 + 8443 + 7559) * 2) + 73{,}728 + 54}{128 * 128 * 3 * 30} = 0.0557 \qquad (4.1)$$

In Eq. (4.1), the numerator values are obtained after obtaining final transmission coefficients after the process of run-length coding, motion vectors estimation while denominator value is given the size of video signal 128 × 128 with 1-s duration for 30 frames. The average values of subjective quality matrices for foreman video signal are summarized in Table 4.1, while values of quality matrices for each frame are shown in Fig. 4.5.

Fig. 4.3 Original foreman video frames

4.3.2   Results for Vipman Video Signal

The original foreman video frames and compressed foreman video frames are shown in Figs. 4.6 and 4.7, respectively. Here, data rate of video signal is 30 fps.

According to the frame rate in this test is 30 frames per second, so that 30 frames are given in compressed format. The compression ratio for this video signal is calculated using the below equation where compressed data rate is obtained after application of video encoder to the video signal.

Fig. 4.4   Compressed foreman video frames for hybrid video codec

Table 4.1   Quality matrices for compressed foreman video signal for hybrid video codec

PSNR (dB)	MSE	SSIM	MSAD	VQM	Blocking beta		Blurring beta	
					Original frames	Compressed frames	Original frames	Compressed frames
22.25	387.26	0.6266	13.036	5.1864	7.15	11.2	24.52	24.88

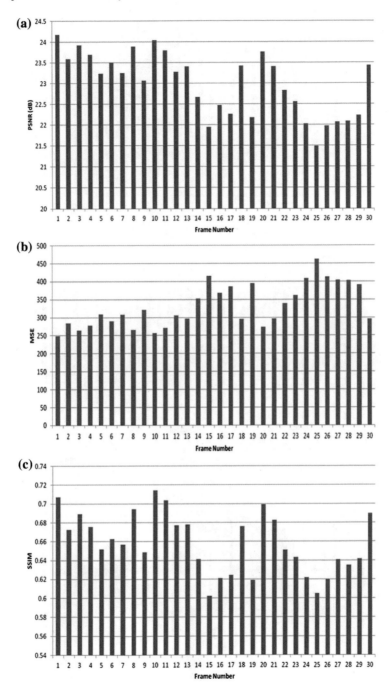

Fig. 4.5 Quality metrics for each frames of foreman video signal for hybrid video codec **a** PSNR values, **b** MSE values, **c** SSIM values, **d** MSAD values, **e** VQM values, **f** blocking beta values, and **g** blurring beta values

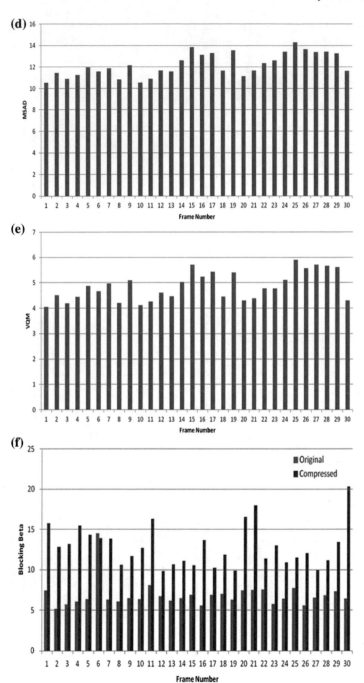

Fig. 4.5 (continued)

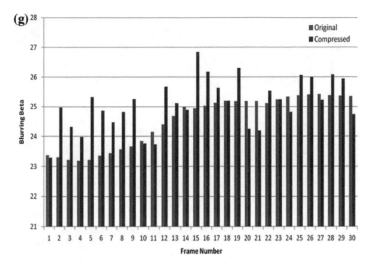

Fig. 4.5 (continued)

$$\text{CR}_{\text{vipman}} = \frac{((4342 + 4872 + 6415) * 2) + 73{,}728 + 54}{128 * 128 * 3 * 30} = 0.0495 \qquad (4.2)$$

In Eq. (4.2), the numerator values are obtained after obtaining final transmission coefficients after the process of run-length coding, motion vectors estimation while denominator value is given the size of video signal 128×128 with 1-s duration for 30 frames. The average values of subjective quality matrices for foreman video signal are summarized in Table 4.2, while values of quality matrices for each frame are shown in Fig. 4.8.

4.4   Summary

The working of standard hybrid video compressor and decompressor is discussed in this chapter. In this chapter, modification in transformation coding blocks by adding two signal transforms such as truncated SVD and DWT along with DCT to improves the compression ratio of the codec. The results of compressed video signals using this hybrid video codec show that the average compression ratio is around 4–5%. The experimental results show that the little decreased is happened in quality metrics such as SSIM, blocking beta, and blurring beta due to high compression ratio. For improving the quality of the compressed video, preprocessing like filtering is applied on compressed video frames which reduce the blocking artifacts and blurring artifacts in it.

Fig. 4.6   Original vipman video frames

Fig. 4.7 Compressed vipman video frames for hybrid video codec

Table 4.2 Quality matrices for compressed vipman video signal for hybrid video codec

PSNR (dB)	MSE	SSIM	MSAD	VQM	Blocking beta		Blurring beta	
					Original frames	Compressed frames	Original frames	Compressed frames
21.667	463.88	0.7553	10.177	5.7058	15.007	8.797	15.757	19.1939

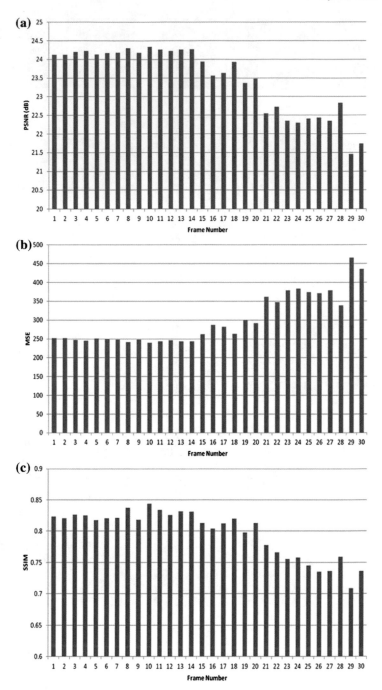

Fig. 4.8 Quality metrics for each frames of vipman video signal for hybrid video codec **a** PSNR values, **b** MSE values, **c** SSIM values, **d** MSAD values, **e** VQM values, **f** blocking beta values, and **g** blurring beta values

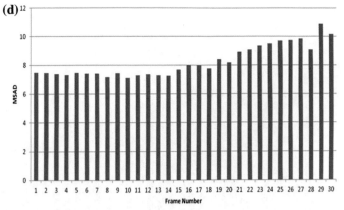

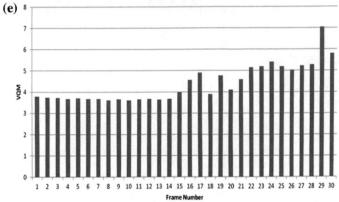

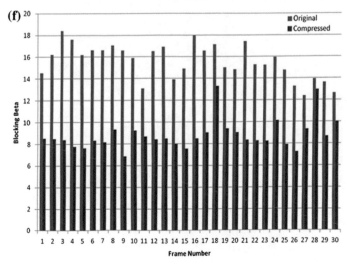

Fig. 4.8 (continued)

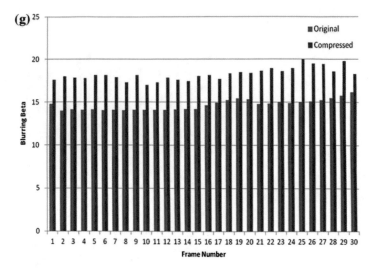

Fig. 4.8 (continued)

References

1. Standard Video Signals. Weblink: https://media.xiph.org/video/derf/. Last Access June 2019.
2. MSU Video Quality Measurement Tool. Weblink: http://compression.ru/video/quality_measure/vqmt_download.html. Last Access June 2019.

Chapter 5
Comparative Comparison of Standard and Hybrid Video Codec

Keywords Standard video codec · Hybrid video codec · Subjective · PSNR

Up to this point, the working of standard video codec and hybrid video codec along with experimental results is discussed in Chaps. 3 and 4. In this chapter, the comparative comparison of these two standards is given with the help of different quality metrics. The comparison is done by using subjective measurement and objective measurement. The parameters such as compression ratio and video quality metrics are used for comparison of standards. Figure 5.1a, b shows the average value of foreman video signal and Vipman video signal for these two standards. The comparison shows that hybrid video codec performs better than standard video codec. The few observations are derived for the comparison of results and given as per below:

- Due to hybridization of transform coding, the compression ratio of the standard increases with the same size of the video signal.
- The blocking artifacts and blurring artifacts are increased in video signal due to compression. But it can be reduced by using hybridization of DWT and SVD instead of using only block-wise DCT-based transform coding.
- The values of PSNR and SSIM are decreased due to hybridization of DWT and SVD, but it is in an acceptable limit.
- The computational complexity of standards video codec does not change by introducing the hybridization of transform coding into it. But it improves the compression ratio of standard.

© The Author(s), under exclusive license to Springer Nature Singapore Pte Ltd. 2020   57
D. R. Bhojani et al., *Hybrid Video Compression Standard*,
SpringerBriefs in Computational Intelligence,
https://doi.org/10.1007/978-981-15-0245-3_5

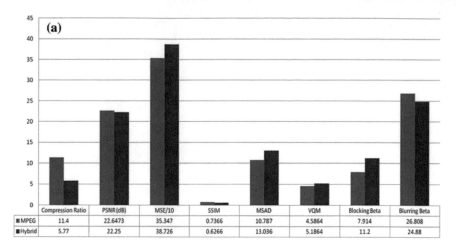

	Compression Ratio	PSNR (dB)	MSE/10	SSIM	MSAD	VQM	Blocking Beta	Blurring Beta
■ MPEG	11.4	22.6473	35.347	0.7366	10.787	4.5864	7.914	26.808
■ Hybrid	5.77	22.25	38.726	0.6266	13.036	5.1864	11.2	24.88

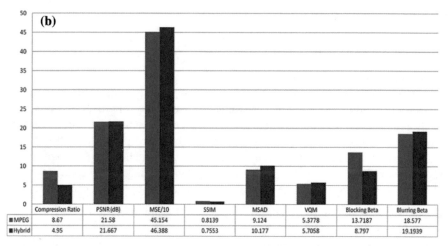

	Compression Ratio	PSNR (dB)	MSE/10	SSIM	MSAD	VQM	Blocking Beta	Blurring Beta
■ MPEG	8.67	21.58	45.154	0.8139	9.124	5.3778	13.7187	18.577
■ Hybrid	4.95	21.667	46.388	0.7553	10.177	5.7058	8.797	19.1939

Fig. 5.1 Comparative comparison of video compression standards **a** for foreman video signal and **b** for vipman video signal

Printed in the United States
By Bookmasters